edZOOcation
Koala Adventures

ZOOKEEPER AGES 3-5

Wildlife Tree
edZOOcation

Meet the cuddly koalas!

Koalas live in Australia. They are fuzzy and cute!

Koalas have a pouch. Baby koalas stay inside.

Can you point to the pouch?

Koalas have soft, gray fur. They have big, round noses.

Can you touch your nose?

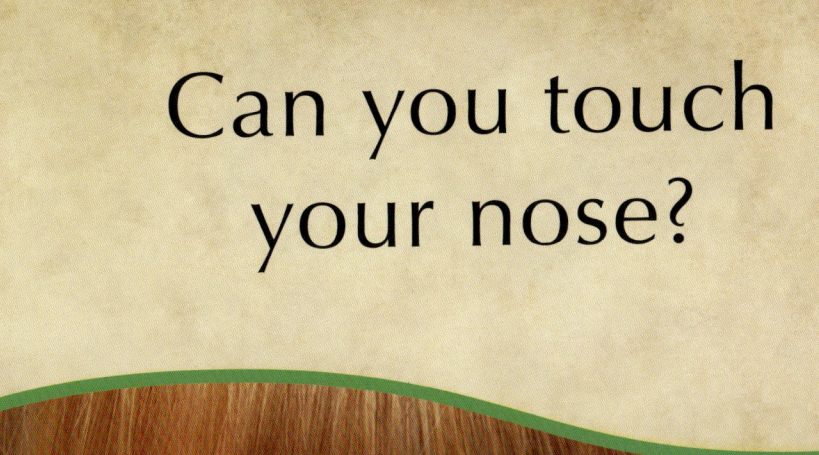

Koalas live in trees. They love eucalyptus trees.

Can you find a tree?

Koalas sleep a lot. They sleep all day long!

Can you pretend to sleep?

Joeys ride on mom's back. They learn to eat leaves.

Can you pretend to ride on a back?

Koalas make funny sounds. They grunt and snore!

Can you make a koala sound?

Koalas live in the tops of trees. They feel safe there.

Can you find a tall tree?

Koalas have two thumbs on each hand!

Can you show your thumbs?

Koalas live with their families. They love each other.

Can you hug like a koala?

Koalas have great senses. They can smell and hear very well.

Can you sniff like a koala?

Goodbye, koalas!

Dedication:

To the curious minds exploring the wonders of our natural world, one koala at a time.

- Jenny Curtis

For Jordan, my sleepy Jellybean.

—A.R.

Curtis, Jenny. Assisted by OpenAI's ChatGPT

Copyright © 2024 Wildlife Tree, LLC. All rights reserved.

Designer: Allyson Randa

Photo Credits:

AdobeStock.com

Pixabay.com

Pexels.com

ISBN: 979-8-9905730-4-8

This book meets **Common Core** and **Next Generation Science Standards.**